Kids & Te Tardigrade Quiz & Fact Book

20 QUESTIONS ABOUT TARDIGRADES

Michael W. Shaw

Fresh Squeezed Publishing
RICHMOND, VIRGINIA

Mike Shaw/Fresh Squeezed Publishing
PO Box 742
Midlothian, VA 23113
www.tardigrade.us

Ordering Information:
Quantity sales. Special discounts are available on quantity purchases by corporations, associations, and others. For details, contact the "Special Sales Department" at the address above.
Tardigrade Quiz & Fact Book/ Michael W. Shaw. —1st Print ed.
ISBN 978-1505977851

Contents

To: Stefanie & Melanie

One touch of nature makes the whole world kin.
—WILIAM SHAKESPEARE

Instructions

You will quickly discover that reading the wrong answers is as much fun as reading the correct answers.

This book does not keep your score. You can go through the book, guessing as best you can at the correct answers, and keep your own score. Then, go through it again and pick all the wrong answers and have fun reading those too!

Why does everyone love guessing games? Maybe it's human nature- from eons ago when we first developed the capacity for language. After people figured out how to ask questions, I imagine that observations of nature probably encouraged the first guessing games. These guessing games were precursors to board games, then centuries later came electronic games, and now, e-books and apps have come along. This edition is the good old paperback book. You can write in it and take notes.

Have fun! Learn some things. That's what counts.

Habitat

Tardigrades live in which type of habitat?

A) The sea

B) Fresh water rivers and streams

C) On land

D) All of the above

ANSWERS:

A) The Sea. True - but not the best answer...

B) Fresh water rivers and streams. True - but not the best answer...

C) On land. True - but not the best answer. Just like all bears, water bears can live on land too, but tardigrades must have moisture to move around in. After a rain, when leaves are moist, and when lichen and moss are wet, tardigrades revive and come out of cryptobiosis (more about that later).

Photo taken at Maymont Park, Richmond, VA

D) All of the above. Correct. Tardigrades live in all environments. Tardigrades can live in all environments where there are other life forms. Obviously, a tardigrade cannot live inside a piece of rock.

Typically, tardigrades can be found in lichen and mosses, which have a tendency to collect moisture, in rooted aquatic plants (which grow in the water), in wet sand, and even in the barnacles on the sides of ships. Tardigrades need a thin film of liquid to move around in because they are aquatic animals, meaning they need to live in water to move, eat, and reproduce.

Tardigrades thrive in all habitats where plants and other animals are found.

Photo taken at Crabtree Falls, Virginia

Nervous System

Do tardigrades have a nervous system?

☐ Yes

☐ No

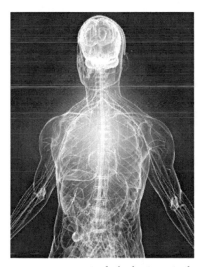

Hint: The human nervous system includes brain, spinal cord, and nerves.

ANSWERS:

☐ Yes. Correct. Just like humans and many other creatures, tardigrades have a nervous system and brain. The tardigrade nervous system consists of the "ganglion," which runs under the belly up to the brain. The brain is part of the nervous system and regulates all the functions of the tardigrade.

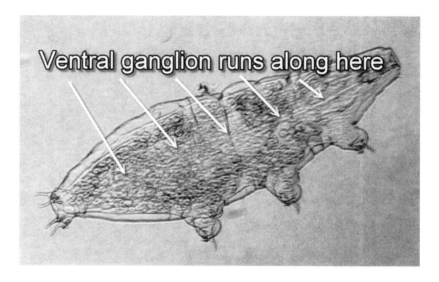

☐ No. Incorrect. Tardigrades do indeed have a nervous system.

Food

W̄hat do tardigrades eat?

A) Bacteria

B) Algae

C) Plants

D) Microscopic animals

E) All of the above

ANSWERS:

A) Bacteria: True, but not the whole answer. Tardigrades do eat bacteria, which are very small. Continue with another guess.

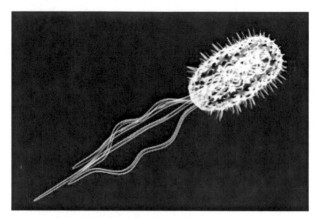

Bacteria Image courtesy of jscreationzs / FreeDigitalPhotos.net

B) Algae: True, but not the whole answer. Yes, tardigrades eat algae. But there is more... Try again.

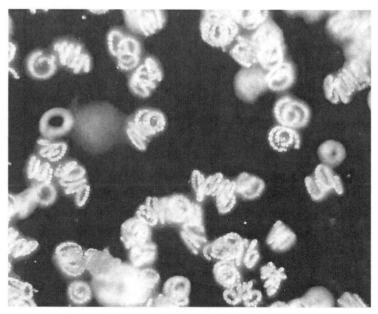

This type of algae is called Spirulina

C) Plants: True, but not the whole answer. Tardigrades use their sharp stylets to pierce the layers of plants and other animals that they eat.

Here is a picture of a plant and a protozoan called Actinosphaerium. This protozoan has spikes coming out of it, probably as a protection and to also help it move around and stick to surfaces.

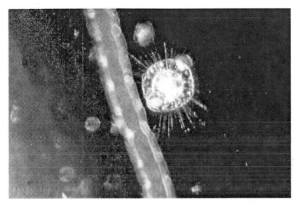

Tardigrade Food

D) Microscopic Animals: True, but not the whole answer. Tardigrades use their sharp "stylets" to piece the skin of animals as they suck out the juices. Some tardigrades also eat other tardigrades. Yuck.

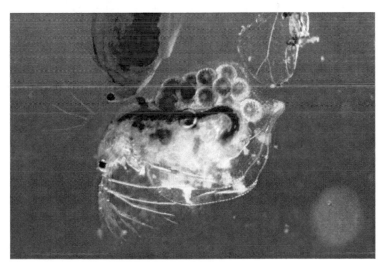

This microcritter is a Daphnia

E) All of the Above: Correct. Tardigrades eat just about everything. Terrestrial tardigrades (those that live on land) eat bacteria, algae, and plants. Tardigrades also eat dead plant and animal fragments (detritus).

Marine tardigrades (those that live under the sea) act as parasites, living on sand dollars, sea cucumbers, algae, barnacles and mussels for their nourishment.

Would you like to do a science project? Find out what tardigrades eat. That would be a good project.

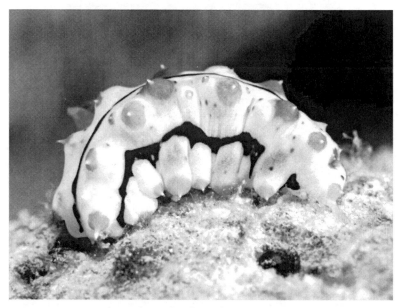

Sea Cucumber

Legs

How many many legs does a tardigrade have?

A) Two (2)
B) Four (4)
C) Six (6)
D) Eight (8)

Learn the Tardigrade Chant:

Two, four, six, eight, who do we appreciate?
Tardigrades,
Tardigrades,
Tardigrades! Yay!

ANSWERS:

A) Two (2) Two legs? No. But that brings up a very famous riddle. It is called the riddle of The Sphinx.

The Sphinx in Egypt asked this riddle.

"What walks on four legs in the morning, two legs at noon, and three three legs in the evening?"

Those who could not answer the riddle were killed and eaten by this mythical beast.

You will find out the answer to this riddle, only when you guess how many legs a tardigrade has.

Photo of The Great Sphinx of Giza, by Sophie Secula

B) Four (4) No. A dog has 4 legs.

So this dog walks into a saloon in the Old West and says "I'm looking for the man who shot my paw."

C) Six (6) Nope. Guess again...

Which creatures do have 6 legs? All insects have six legs.

Well, what about the spider- isn't a spider an insect? Technically, no. A spider is not an insect because it has the same number of legs as a tardigrade (hint).

Spiders are arthropods just as insects are, but they differ in their number of body segments, and they do not have antennae. They are called arachnids, in the family of mites and ticks.

Spider Image courtesy of panuruangjan / FreeDigitalPhotos.net

D) Eight (8) Correct. 8 legs.

That's right - a tardigrade has 8 legs (4 sets - each with 2 legs).

The first three pairs of legs face forward, and the last set of legs faces backwards. Why? The back legs are primarily used to hold onto something for stability, while the front three pairs are used for crawling.

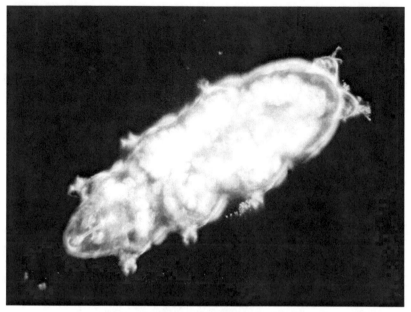

Notice the rear set of legs facing backwards

And now the answer to the Sphinx riddle:

"What walks on four legs in the morning, two legs at noon, and three three legs in the evening?"

Answer: A man. Why? As a baby it crawls on four, as an adult walks on two, and as an elderly person, uses a cane.

Experiment

Eleventh graders from Rochester, New York did what kind of experiment with tardigrades?

A) They sent tardigrades up in a hot air balloon

B) They built a small rocket and sent the tardigrades high up into the atmosphere

C) They sent the tardigrades into space, up to the International Space Station

ANSWERS:

A) Hot Air Balloon: Incorrect. How would you fit the whole class and the teacher in the basket under the hot air balloon? That would have been a lot of fun trying though.

Balloonfest in Statesville, North Carolina

B) Small Rocket: Wrong. Model rocket too small. Would not get out of Earth's atmosphere.

Think big...

C) Space Station: Correct. Good guess. The experiment was to see if the tardigrades would dehydrate and come back to life in zero gravity. They did!

The International Space Station (ISS) is a station that orbits Earth, about 200 miles (322 km) above the surface. Crew astronauts live on the space station for months at a time and perform experiments. If you look up in the sky on a very clear night, you can actually see the ISS orbiting, lit up like a tiny star. You have to know when and where to look. Go to the "Spot The Station" NASA website and type in your location to find out where to look.

Space Shuttle and International Space Station

Suspended Animation

When tardigrades go into a state of suspended animation this is called:

A) Zombie

B) Kryptonite

C) Cryptobiosis

ANSWERS:

A) Zombie: Incorrect. In order to become zombies they'd have to be dead and then move around. Tardigrades don't do that because they are too small and too proud to star in silly movies.

Tardigrades do not die and come back from the dead. They don't go into a crypt; they go into cryptobiosis.

Actual sign

B) Kryptonite: Wrong Answer. Tardigrades may be super creatures, but they don't go into suspended animation when near the real element called Krypton. The element Krypton has the symbol Kr, and an atomic number of 36. It is a colorless, odorless, gas. It is used in lighting and photography in fluorescent lamps.

Tardigrades Welcome

C) Cryptobiosis: Correct. Very good. Now practice saying it. KRYPTO-bye-oh-sis.

When a tardigrade is dehydrated or has its environment threatened in other ways, like freezing or extreme heat, it curls up into a little barrel shaped ball called a "tun."

The water inside the tardigrade is replaced with a type of sugar called "trehalose," which allows the tardigrade to stay preserved like this for long periods of time, even many years. Eventually, the tardigrade will come back to life, perfectly normal when favorable conditions occur again.

More about this process later...

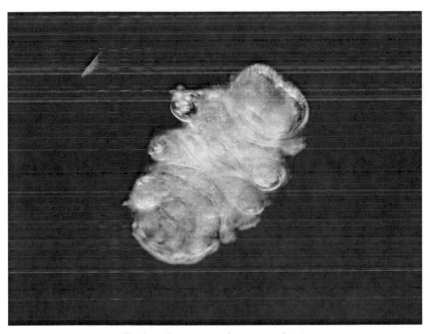

Tardigrade emerging from cryptobiosis

DNA

D o Tardigrades have DNA?

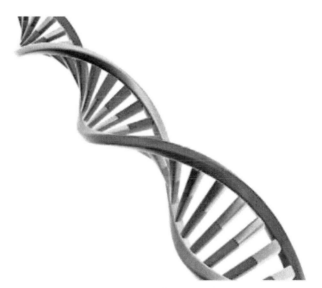

Image courtesy of CoolDesign / FreeDigitalPhotos.net

☐ Yes

☐ No

ANSWERS:

☐ Yes: Correct. They do have DNA. Anything that is alive has DNA. Plants, animals, bacteria, fish, tardigrades all have DNA.

DNA is the basic structure of life, and found in just about every cell or every living thing. Obviously super tiny, it is located in the nucleus of the cell. DNA contains information telling each cell what to do, for example DNA tells a muscle cell to be a muscle cell, or tells a skin cell to be a skin cell.

DNA is a set of instructions for every cell. If this is hard to believe, wait 'till you hear the rest. These instructions are actually in an easy to read code. The code consists of four chemicals that bond in various combinations. Adenine combines with Thymine, and Guanine combines with Cytosine. Let's use the letters A-T and G-C.

These pairs always connect as just explained. A very long series of these pairs becomes a set of simple instructions, like "make blue eyes." It might read like this: A-T, A-T, G-C, A-T, G-C, G-C, and so on, for maybe a list of a few thousand pairs in a long sequence of A-T's and G-C's. Within this section of a strand of DNA this is one instruction, called a gene. All of your genes make you, one human.

All of another group of genes makes a chimpanzee. The genes in a chimpanzee and a human are about 96% the same. Only four per cent makes the difference. Oooh! oooh! I feel like scratching my left ear by putting my right hand over my head. Aaah, that feels good.

Scientist analyzing the DNA of a tardigrade

☐ No: Incorrect answer. All life has DNA, including tardigrades.

Everything in this picture has DNA

Digestion & Muscles

Tardigrades have a digestive system, and muscles. True or False?

☐ True

☐ False

ANSWERS:

☐ True: Correct. Just like humans and many other creatures, tardigrades can eat and digest food, and they use muscles to do this.

Tardigrades use muscles of their"pharynx" (between the mouth and stomach) to move the food from mouth to stomach. In the picture, you can actually see this mass of muscle tissue in the pharynx.

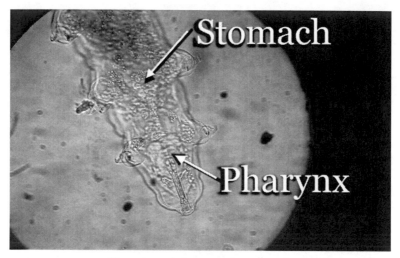

Trace the path from mouth to stomach

☐ False: Incorrect answer. Tardigrades do eat a lot. They need a digestive system to do it.

Tardigrades have to eat too!

Age

The oldest specimens of tardigrades ever found are:

A) 90 million years old

B) 8 years old

C) 153

ANSWERS:

A) 90 million years old. Correct. A tardigrade was discovered perfectly preserved in a piece of amber about 90 million years old found in New Jersey, and a tardigrade was also found in Canadian amber of about the same age.

Tardigrades have not been found in fossilized rock, because they are too soft. Only bones and hard materials like shells are found in fossilized rock. Amber is different because it is simply sap from a tree, like maple syrup. Many insects, different types of pollen, small micro-organisms are easily trapped in this gooey sap.

Then, over millions of years this sap becomes hard, and eventually turns into the stone we call amber. Inside, we can see perfectly preserved specimens, and that's how we discovered tardigrades in amber.

In the below photo, you will see an ordinary 90 million year old ant, perfectly embedded. Do you have any amber jewelry in your house? Look inside with a magnifying glass. You might discover a tardigrade.

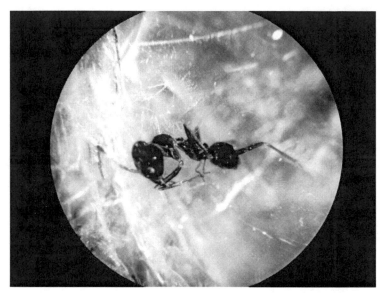

This is an ant, preserved in amber, about 90 million years old.

B) 8 years is incorrect. I have tardigrades older than that in the dirt on my old boots.

Old Boots Image courtesy of artur84 / FreeDigitalPhotos.net

C) 153. Incorrect. That's the number I always guess when there is a prize for guessing how many jelly beans are in a jar.

Cryptobiosis

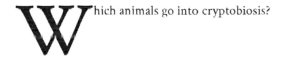

W hich animals go into cryptobiosis?

A) Rotifers

B) Nematodes

C) Tardigrades

D) All of the above.

ANSWERS:

A) Rotifers. Yes. But there is a better answer. Rotifers are small micro animals that live mainly in freshwater environments. If you take a sample of pond water and look at it on a microscope slide, you are likely to find rotifers.

Some rotifers swim, and others creep along a surface like an inchworm. They can go into cryptobiosis just like tardigrades. Rotifers are a major food source for many microorganisms.

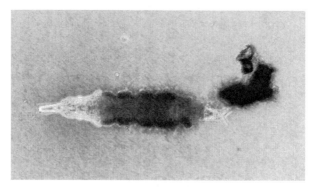

Rotifer eating a piece of microscopic plant life

B) Nematodes. Yes. But there is a better answer. Nematodes are round-worms, and there are over 25,000 known types.

Shown here is a microscopic nematode, often eaten by tardigrades. Some nematodes, like the one shown here, can go into cryptobiosis just like tardi-grades. They have a tubular digestive system with openings on both ends. Nem-atodes are a major food source for many microorganisms.

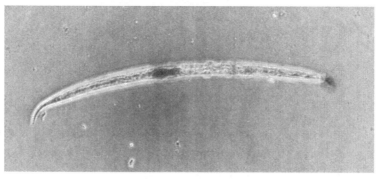

Microscopic Nematode

C) Tardigrades. Yes. You're getting warm... Of course tardigrades can go into cryptobiosis.

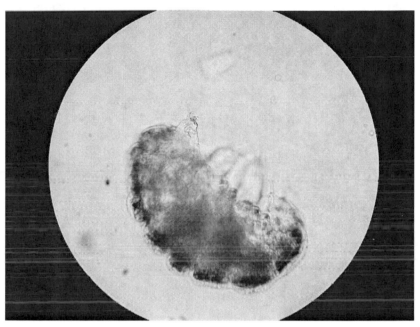

Tardigrade going into cryptobiosis

D) All of the above. Yes. All of these animals go into cryptobiosis.

Rotifers and nematodes as well as tardigrades can perform this trick of suspended animation. Cryptobiosis is explained in detail in the next question.

Can a human go into cryptobiosis?

The only human to have done it was Rip Van Winkle. Rip Van Winkle is a story about a man who goes to sleep for about 20 years. The story is in the author Washington Irving's: The Sketch Book of Geoffrey Crayon, Gent.

Bronze of Rip Van Winkle sculpted by Richard Masloski, copyright 2000

Stages

When tardigrades go into cryptobiosis they form a:

A) Spore

B) Egg

C) Cyst

D) Bun

E) Tun

ANSWERS:

A) Spore. Incorrect. A spore is a little seed that becomes a mushroom, fungus, or other plant, and some microscopic animals.

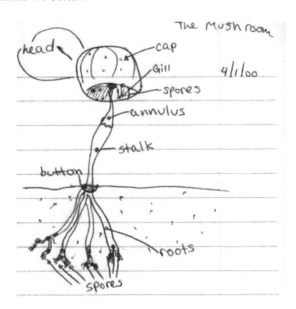

Drawing by Stefanie Shaw. Spores are under the cap and at the roots.

B) Egg. Incorrect. An egg is something you find on an English muffin with a layer of American cheese on top and a slice of Canadian bacon on the bottom. This is because the English hatched both Canada and the United States.

By the way, do you know the muffin man?

Hey. Who makes the best breakfast?

C) Cyst. Incorrect. A cyst is an abnormal growth, or sometimes a protective encasement that a parasite lives in.

If you put dirt in your mouth, even accidentally, you could ingest round-worm eggs. These will hatch in your body, and become parasites in your tissue. The eggs live in cysts waiting to explode into the parasites.

Don't eat dirt. Wash your hands before eating. This is a good reason.

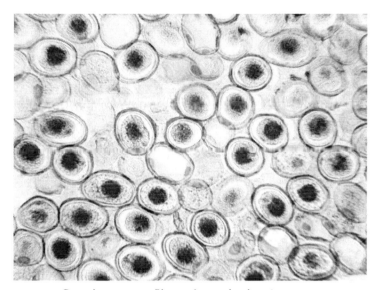

Roundworm cysts. Photo taken under the microscope.

D) Bun. Incorrect. A freshly baked cinnamon bun would be a nice snack right now. Want the recipe? Buy a refrigerated tube of cinnamon buns. When you unroll it, string them together into one big coil, instead of making individual buns.

Place in a cast iron or small greased round baking pan. Bake it at 350 F covered with foil for about twice as long as it says on the package- because it is more dough and needs to bake longer. Remove the foil during the last 10 minutes to brown the top.

Remove from oven using pot holders, and coat with the creamy frosting that comes in the package.

Baked in a small cast iron pan, and yummy!

E) Tun. Yes! Correct. This is a state where all tardigrade metabolism, all body activity, has stopped. Tardigrades become resistant to heat, cold, chemicals, drought, and even radiation.

How does it work?

When water freezes, it expands. That would kill any living organism whose cells have water, because all the cells will burst. The tardigrade is different because as it dries out or freezes, the water is replaced with a sugar called "trehalose."

In a tardigrade, the glucose (normal sugar like ours) is replaced with trehalose sugar. This preserves the tardigrade because trehalose forms a protective barrier, a kind of a gel, that prevents expansion of any remaining fluids.

Humans and other animals do not have that ability. We are still studying how this process works.

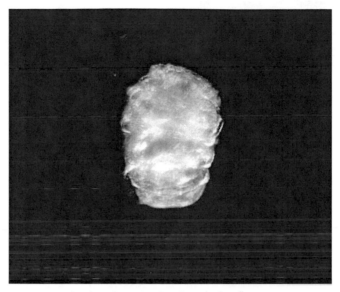

Tardigrade contracting into the "tun" stage.

Body Parts

Can you identify these parts of the tardigrade?

A) Brain

B) Salivary Gland

C) Tubular Pharynx

D) Sucking Pharynx

E) Mouth

F) Ventral Ganglion

G) Stylet

H) Esophagus

I) Ovary

J) Eggs

K) Eyespot

L) Claws

M) Cloaca

N) Oviduct

O) Stomach

ANSWERS:

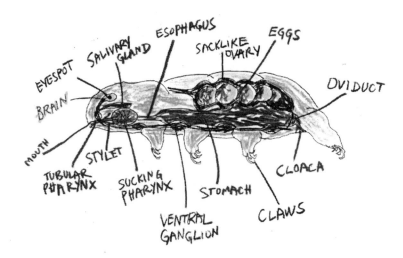

A) Brain – Regulates all activity of the tardigrade. Attached to the nervous system called ganglia.

B) Salivary Gland – This controls the moisture to the mouth of the tardigrade.

C) Tubular Pharynx - This is actually part of the mouth that is elongated into a tube to the stomach.

D) Sucking Pharynx - This is a clearly visible organ that is a mass of muscles which suck the food using vacuum before it goes into the stomach.

E) Mouth - Tardigrade mouths come in a variety of interesting shapes, and that is one way to tell the various tardigrade species apart. The mouth is used for sucking food, so you will see it shaped like a round sucking device.

F) Ventral Ganglion - Ganglia are nerves. In humans our ganglia would be in the spinal cord running up our backs. This is considered "dorsal" meaning toward the back. "Ventral" means the belly. The tardigrade has its nervous system running along the belly.

G) Stylet - these are sharp pointers that are used for piercing harder surfaces of animals and plants so the tardigrade can suck out the softer contents.

H) Esophagus – The canal leading to the stomach from the mouth.

I) Ovary - Contains the eggs until they mature.

J) Eggs - Reproductive containers for tardigrade embryos. Tardigrades lay eggs, but some do not, and the tardigrade sheds its husk with the eggs inside. The husk, or outer skin of the tardigrade keeps the eggs together until they all hatch, and this possibly protects the eggs from predators.

K) Eyespot - The tardigrade eye is not complex. It is a very simple light sensitive spot, that allows the tardigrade to sense brightness or dark.

L) Claws - Tardigrade claws are very distinctive between species. This is the easiest way to tell various types of tardigrades apart. Some have one long claw and three short claws on a foot. Others have two long and two short, and these claws have different shapes and curves that make it really easy to tell apart the different species.

M) Cloaca Rear port for excretion and reproductive functions.

N) Oviduct - part of the egg reproductive system.

O) Stomach – Also known as the tummy, where food goes that is yummy.

Size

How big is a tardigrade?

A) 2 inches

B) 5 feet 7 inches

C) About half the size of a cat

D) 0.1 to 1.2 mm

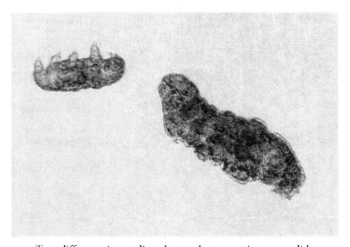

Two different size tardigrades on the same microscope slide

ANSWERS:

A).2 inches. Seriously? Nope. But a 2 inch tardigrade would make a cute pet. Take a look.

Awww. Isn't he cute?

B) 5 feet 7 Inches. Almost right. If it were an inflatable beach toy.

Beware of giant tardigrades. Most likely seen in marine environments.

C) About half the size of a cat. Correct, if it is a stuffed animal tardigrade.

My cat and his tardigrade cat toy

D) 0.1 to 1.2 mm. Correct. The largest tardigrades are about 1mm

And in microscopic terms - that is gigantic. Bacteria, or a blood cell is a thousand times smaller. Most microscopic life in a pond, for example, is about 1/4 the size of a tardigrade.

Take a look at this picture of a 1 millimeter rule- under the microscope. It is divided into 100 sections. Remember - the whole thing is only 1 mm.

That's how we measure tardigrades. We put them on a ruler like this one. The biggest divisions are 1/10 of a mm, or shown in the picture as .1mm - that's one tenth (1/10 as a fraction) or 0.1 in decimal form.

The smallest tardigrade is .1mm. the largest is 1.2 mm, and would overlap the ruler. Remember this ruler is 1mm long.

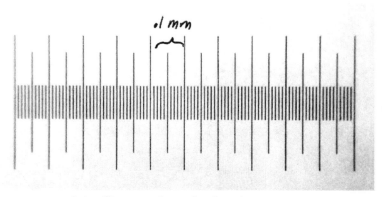

A 1 millimeter ruler under the microscope

Continents

Tardigrades have been found on how many continents?

A) 4 continents: Afro-Eurasia, America, Anarctica, Australia

B) 5 continents: Africa, Eurasia, Antarctica, Australia

C) 6 continents: Africa, Europe, Asia, America, Antarctica, Australia

D) 7 continents: Africa, Europe, Asia, North America, South America, Antarctica, Australia

ANSWERS:

A) True. Tardigrades are found on 4 continents, but guess again. It depends how you count the continents. In this method, both North and South America are counted as one continent, and Africa, Europe and Asia are counted as one continent. That leaves Antarctica and Australia as the remaining continents, giving a total of 4 continents. Since tardigrades are found on all continents, your answer is technically correct, but guess again...

Fun facts: On the African continent, between 2000 and 3000 different languages are spoken in 54 different countries. The Australian continent has one country: Australia. Australia has no official language, however English is the primary language spoken there.

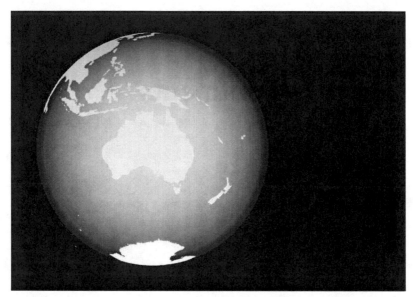

Australia and Antarctica

B)   5 continents. True, but again, it depends how you count the continents. In this method, both North and South America are counted as one continent, and Africa is separated from Europe and Asia (which are combined as one continent). That leaves Antarctica and Australia as the remaining continents, giving a total of 5 continents. Remember, tardigrades are found on all continents, so your answer is technically correct, but guess again...

Fun Facts: Europe has 50 countries, plus 7 that are disputed. There are about 140 languages commonly spoken there. There are 14 countries in South America. Most have Spanish as their main language except Brazil where the population speaks Portuguese and the Falkland Islands where they speak English.

Counting continents can be confusing

C) 6 continents. True, and we are so close, but not there yet. Here's another method to count the continents. In this method, both North and South America are counted as one continent, and Africa, Europe and Asia are counted each separately. That leaves Antarctica and Australia as the remaining continents, giving a total of 6 continents. Technically, you are correct because tardigrades are found on all continents. See the correct answer in your next guess.

Fun Facts: The North American continent has 3 countries - Mexico, U.S.A., and Canada. There are 48 contiguous United States joined together on the North American continent, and then Alaska for state number 49, and finally we have Hawaii, the 50th state, a series of islands in the Pacific Ocean. The official language of the U.S.A. is English, the official languages of Canada are French and English, and the official language of Mexico is Spanish. Compare this to the Asian continent: There are 49 countries plus 5 disputed countries in Asia. Asia has about 2269 languages.

D) 7 continents: True. This and all the above all answers are correct. If you count the continents as 7, you are using this method. Both North and South America are counted individually as two continents. Africa, Europe and Asia are counted each separately. That leaves Antarctica and Australia as the remaining continents, giving a total of 7 continents. As we've said before, tardigrades are found on all continents, so all answers are correct.

Did you know that the continents, however you wish to count them, were all one big continent? About 300 million years ago, during the late Paleozoic era, all of the continents were one big land mass, which we call Pangaea. Over time, this super continent split up and shifted apart by continental drift into what we now have as separate land masses or continents.

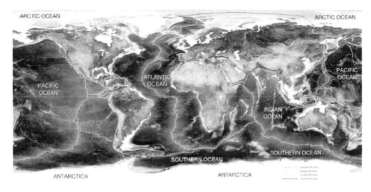

Long ago, the continents were all one.

Fun Fact: Antarctica - Tardigrades have been found in Antarctica, under extreme conditions, there are limited places for them to move around and obtain food. There are seven countries that claim territory in Antarctica: United Kingdom, New Zealand, France, Norway, Australia, Chile, Argentina. These territories are agreed upon by treaty between the members.

Do penguins live in Antarctica? Some do, but these are African Penguins.
12 of 17 species of penguin live in warm climates.

Species

How many species of tardigrades have been discovered so far?

A) Under 100
B) Slightly over 200
C) In the range of 1000

ANSWERS:

A) Under 100. Incorrect. Guess again. As a comparison, there are 78 species of whales; only 11 of these species have teeth.

If tardigrades were whales, you would have been right, but whales are very big, and tardigrades are small. Very, very small.

Beluga Whales Image courtesy of Victor Habbick / FreeDigitalPhotos.net

B) Slightly over 200. Incorrect. As a comparison, there are between 230 and 270 species of primates, one of these species being human.

We are primates, you know, the ones that talk too much, and like to wear warm hats and pose for pictures with other primates. The other couple of hundred species of primates are normal.

Two primates

C) In the range of 1000. Yes correct. About 1000 species of tardigrades have been found.

We are discovering new species of tardigrade all the time. As a comparison, there are 10,000 species of birds.

The tardigrade requires a particular habitat that alternates between wet and dry, or one that is always wet. So it is not likely that there would be so many species of tardigrade as we have of birds.

The hummingbird is the smallest bird weighing a few ounces and with a wingspan of a few inches. The smallest hummingbird weighs less than a penny. The ostrich is the heaviest bird weighing 240 pounds (109 kgs), and the Golden Eagle has a wingspan of over 8 feet (2.4 meters).

Those are just three examples of bird diversity, and there are 10,000 more types.

Of the thousand or so species of tardigrade, which is not that many, we are not likely to find the diversity we find in birds.

Hummingbird photo taken in Juliette, Georgia. Weight: about 1 ounce.

Baby Ostrich with two eggs. Weight about 10 Lbs.

QUESTION 16

Habitat

Terrestrial tardigrades (they live on land) are found in what types of habitats?

A) Leaf litter
B) Mosses
C) Lichen
D) Rocky beach
E) Sand
F) Arctic ice
G) Tree bark
H) All of the above

ANSWERS:

A) Leaf litter. Not the best answer - try again. Yes, tardigrades are found in moist leaf litter. Try again though.

When leaves become wet, tardigrades are up and about.

B) Mosses. Not the best answer - try again. Yes, tardigrades are found in mosses. There is a better answer. Try again.

Moss

C) Lichen. Not the best answer - try again. Yes, tardigrades are found in lichen.

Lichen gets all of its nutrients from the air around it. If there are pollutants in the air, it absorbs those too. Can tardigrades teach us about pollution?

A tree covered in different types of lichen

D) Rocky beach. True. But there is a better answer. Tardigrades have been found in mussels and barnacles on rock sea shores.

E) Sand. Correct. But there is a better answer. Tardigrades are found in beach sand, but there is better answer. Try again.

Outer Banks, North Carolina. Photo by Donna Shaw

F) Arctic ice. Correct. But there is a better answer. Tardigrades have been found in the arctic and in Antarctica. There is a better answer, so try again.

In arctic climates we find tardigrades living in the mosses and lichens on the land, in soil and sand, in ice, and we find them in sea barnacles and in freshwater areas.

Antarctica on a warm day

G) Tree bark. Correct. But there is a better answer. Tardigrades are found in tree bark, but there is a better answer.

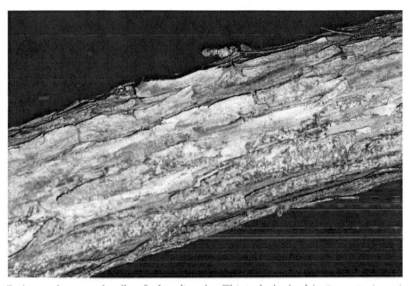

Bark must be scraped well to find tardigrades. This is the bark of the Dawn Redwood.

H) All of the above. Correct - this is the best answer. Tardigrades are found in all of the above habitats. The easiest places to find them, in any environment, are in lichen or in moss. Tardigrades need a thin film of liquid to move around in because they are aquatic animals, meaning they need to live in water. Even though they may live on land, they still need to be kept wet.

Lichen covered tree - loaded with tardigrades

Respiratory & Circulatory

Tardigrades have a respiratory system and circulatory system. True or False?

☐ True

☐ False

ANSWERS:

☐ True: Not correct. Tardigrades do NOT have blood circulating throughout their bodies. And they do not have lungs that breathe.

☐ False: This is the answer. Why? To circulate the fluids in their body, tardigrades must use their entire body by expanding or contracting it. Like other animals of this type, without lungs, they take in oxygen through their skin, also called the cuticle. They have no blood and no lungs.

In humans and other animals, blood and lungs (circulatory and respiratory) are really connected. The respiratory system which we humans have (our breathing), feeds oxygen into the blood through little air sacs in our lungs. The blood in our circulatory system runs through the capillaries in the lungs, and absorbs oxygen. This is all one machine.

Oxygen then goes to the brain keeping us alive and able to think and allows the brain to control the rest of the body. Anything that happens to the body to prevent oxygen from getting to the lungs is very harmful to us.

Since tardigrades do not have a respiratory system (lungs) nor a circulatory system (blood), they can survive more extreme conditions than we could survive.

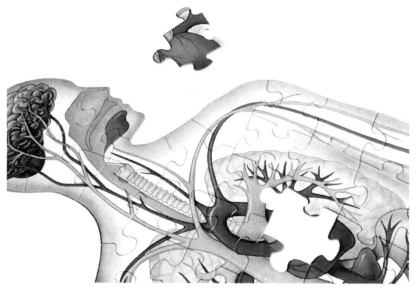

Heart and Lungs work together to deliver Oxygen to the Brain

Extremes

Tardigrades can withstand which of the following extremes?

A) -200 Degrees Centigrade (- 328 Fahrenheit)

B) +120 Degrees Fahrenheit (49 Degrees Centigrade)

C) 27,000 PSI (pounds per square inch of pressure)

D) Harmful ultraviolet light

E) Harmful X-Rays

F) The extreme vacuum of space

G) All of the above

ANSWERS:

A) -200 Degrees Centigrade (- 328 Fahrenheit). Correct. But there is a better answer. By the way, -200 Degrees Centigrade (- 328 Fahrenheit) is pretty cold, and by comparison, the arctic is warm. The coldest temperature recorded in the arctic is - 68 C (-90F).

In order to test how cold a temperature a tardigrade can withstand, you need to put it under that left over quart of ice cream in the back of your freezer. On second thought, just eat the ice cream.

Dry ice is -78 C (-109.3 F) but tardigrades can handle colder than that.

Liquid nitrogen at the surface is about -196 C (-320 F), so a science teacher or professor might be able to test if a tardigrade would survive in that.

Liquid nitrogen. Now that's cold.

B) +120 Degrees Fahrenheit (49 Degrees Centigrade). Correct. But there is a better answer. In case you are wondering, +120 Degrees Fahrenheit (49 Degrees Centigrade) is pretty hot, but not that hot. No, it's not nearly as hot as a volcano. The lava in a volcano is at least 700 C (1292 F). Even a tardigrade would not survive in a temperature hot enough to melt rock. That's hot.

Lava from a volcano

C) 27,000 PSI (pounds per square inch of pressure). Correct. But there is a better answer. What this PSI all about? 27,000 PSI is more pressure than under the deepest part of the ocean. And that's a lot. So tardigrades are cool for that reason alone. The deepest part of the ocean is called the Marianas Trench, located in the Pacific Ocean. It reaches a depth of 6.831 miles (10.91 km). The pressure is 15,750 psi (1086 bars). Researchers have reported microscopic organisms living in the Marianas Trench.

Tardigrades can survive under even more pressure. Let's take a lesson from that and not complain so much when we feel like we're under pressure.

"No pressure, little guys, but I think I see a shark coming."

D) Harmful ultraviolet light. Correct. But there is a better answer. Don't be like a tardigrade! Ultraviolet light, even from a sunny day on the beach can be harmful. Always wear a sunblock, and sunglasses. Wear a hat to prevent sunstroke. If you wear sunblock at the beginning of summer, you will still get a tan, if that's what you want.

Tardigrades have been subjected to ultraviolet light 1000's of times stronger than you would get on the beach, and they have survived. You are not a tardigrade!

E) Harmful X-Rays. Correct. But there is a better answer. Let's put it this way. If a tardigrade went to the dentist, the dental assistant wouldn't have to use one of the big lead apron things to protect the tardigrade from X-rays. The hard part would be getting the tardigrade to use dental floss regularly.

F) The extreme vacuum of space. Correct. But there is a better answer. Did you know that tardigrades have been taken up into space and exposed to the vacuum? Meanwhile the astronauts had to wear space suits. Ha!

Astronaut in the extreme vacuum of space

G) All of the above. Yes. That is the best answer. Tardigrades can withstand all of these extremes! Tardigrades have been voted the most extreme animal on the planet.

Tardigrade - The Most Extreme Animal on the Planet

Eggs

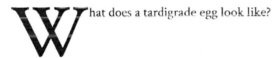

What does a tardigrade egg look like?

A) A ball with spikes

B) A ball with spiky spikes

C) A ball with very spikey spikes

D) A ball with thumb tacks stuck in it

E) A ball with pushpins stuck in it

F) A smooth round ball

G) All of the above

ANSWERS:

A) A ball with spikes. Correct. But there is a better answer. Take a look at this tardigrade egg. Looks just like one of the spikey balls you see in the toy store!

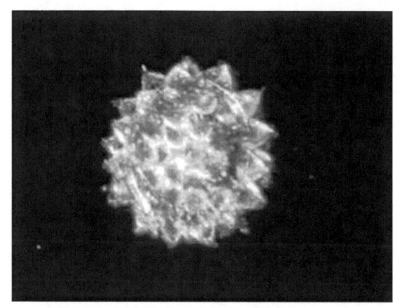

Tardigrade Egg probably from *Macrobiotis species*

B) A ball with spiky spikes. Yes. but there is a better answer.

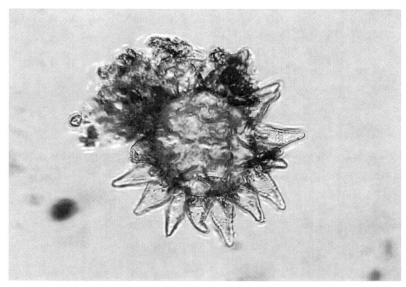

Another type of tardigrade egg probably of *Hypsibius species*

C) A ball with very spikey spikes. Yes. But there is a better answer

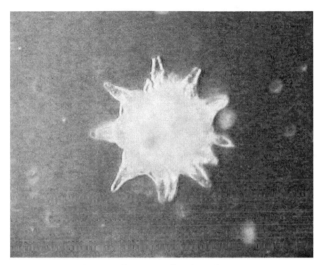

Yet another type of tardigrade egg

D) A ball with thumb tacks stuck in it. Correct. But there is a better answer.

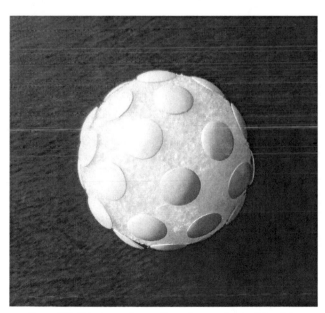

Yes - *Minibiotis species* tardigrade eggs looks like this.

E) A ball with pushpins stuck in it. Correct. But there is a better answer.

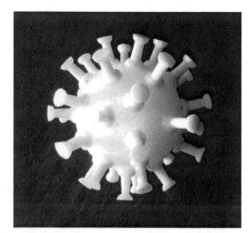

Yes -*Macrobiotis species* has an egg that looks like this

F) A smooth round ball. Correct. But there is a better answer. These oval, smooth eggs are found inside the empty husk, also called a cuticle, which the female leaves behind. The eggs hatch and break open the husk.

Tardigrades have a series of the husks which they outgrow, and shed, like a snake sheds its skin. Each successive husk is called an "instar." It is also sometimes called molting.

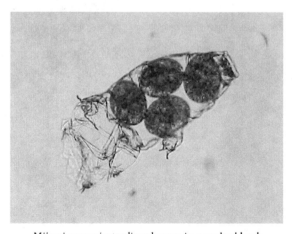

Milnesium species tardigrade eggs in a molted husk

G) All of the above. Correct. This is the best answer. Tardigrade eggs come in various strange shapes, smooth, spiky and weird. You can make your own replicas from things you will find in the crafts store. Did you ever see a ball in the toy section of the store with spikes on it? That looks like a big tardigrade egg.

Once the egg hatches, tardigrades grow like other animals, getting bigger over time. They outgrow their skin, and shed it like a shell, and do this again and again. They can do this from 6 to 12 times from baby to adult.

All of these look like tardigrade eggs

Space

Tardigrades came from outer space. True or False?

Asteroid impact - Image courtesy of Idea Go / FreeDigitalPhotos.net

☐ True

☐ False

ANSWERS:

☐ True: Not correct. This cannot be correct. Find out why by trying again.

Wait - before you go to the correct answer - one more quick question:

Which is tastier - a Moon rock or an Earth rock?

A Moon rock, because it's a little meteor. (meatier)

Okay. Now find out the correct answer whether or not tardigrades came from space.

☐ False. This is the correct answer. Here's why:

DNA analysis of tardigrades shows that they are related to all of the other forms of life on earth. So tardigrades, by themselves, could not have come from outer space.

Did all life on Earth come from outer space?

There is a theory that all life on earth originated in outer space. That theory, called "Panspermia," presents the possibility that some very low life form like bacteria, or even DNA fragments from another planet arrived on earth, and all life evolved from that.

Well, tardigrades would be included in that evolution, so only in that case, if ALL life on Earth originated in outer space, can you say yes, tardigrades, with everything else, came from space.

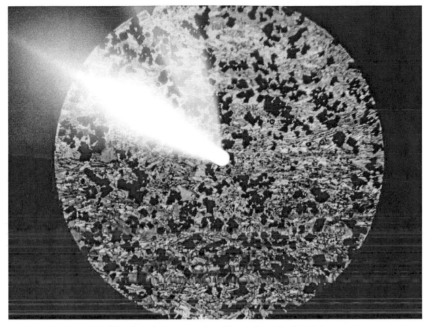

Slice inside a meteorite. No tardigrades inside!

Congratulations!

You made it to the end!

Now – Here is what we do not know about tardigrades:

- How are they distributed? On the wind, on the feet of birds, on dust?
- From where did tardigrades evolve? They seem to come from somewhere between Nematodes and Arthropods.
- How do they find food and each other?
- Are they attracted to light, to heat, to cold, or Oxygen or CO_2?
- What are the effects of pollution on tardigrades?

Would you like to find tardigrades and do a science project about them? Here is a book to help you:

Kids & Teachers Tardigrade Science Project Book

The author's website: www.tardigrade.us

Google this: "First Animal to Survive in Space," and see the author as host, talking about tardigrades.

Check out YouTube: "Songs For Unusual Creatures." See the author host the hunt for tardigrades.

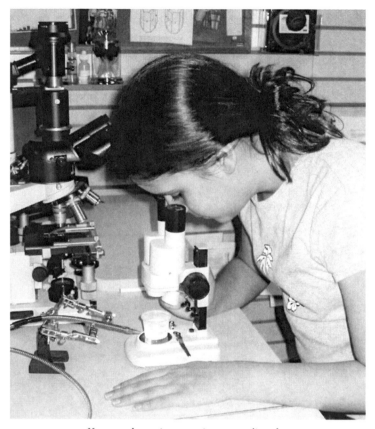

You can do a science project on tardigrades

Will you do the author a favor and review this Kids & Teachers Tardigrade Quiz and Fact Book so he will know if you liked it or not? Just find this book on any site, like Goodreads, Amazon, Smashwords, Barnes & Noble, or Apple iBooks, and write a review. That would be so much appreciated!

Other Books by Michael Shaw:

Your Microscope Hobby - How to Make Multi-Colored Filters

Kids & Teachers Tardigrade Science Project Book

Word Nerd – Things Way Up High Quiz & Fact Book

Websites:

www.tardigrade.us and www.mikeshawtoday.com

www.amazon.com/author/mikeshaw247

http://astore.amazon.com/mikesmicroscopestore-20

Have a great day, have fun, and always learn something new.

ABOUT THE AUTHOR

Michael Shaw took a serious interest in science when his children were in middle school, and he helped them with their science projects. This led to a personal passion to make a contribution to science, and thus began a population survey of the little known creature (at the time), the tardigrade. The survey, covering the state of New Jersey, and publication of a scientific paper took about 10 years from start to finish.

This also resulted in a great deal of how-to knowledge, and a viral internet video popularizing the tardigrade as the "first animal to survive in space." He has starred in the PBS video on YouTube, "Songs For Unusual Creatures," and has appeared on television in Brasil.

Mr. Shaw has written several books to assist teachers and their young scientists in similar projects. Continuing to write, he is currently working on a novel and several inspirational books.